Teach Your Kids Math: Multiplication Times Tables

Copyright & Other Notices

Updates, news & related resources from the author can be found at
http://www.suniltanna.com/multiplication

Information about other math books by the same author can be found at
http://www.suniltanna.com/math

Introduction

For some time now, I have tutored both children and adults in math and science. This book is based on my own personal experience as a tutor, and is one of a series of books on different topics:

- If you want to find out about the other math books that I have written, please visit: http://www.suniltanna.com/math

- For science books that I have written, visit: http://www.suniltanna.com/science

Anyway, let's get started with the topic of this book... teaching your kids multiplication tables.

Knowing multiplication tables is a key math skill. This is because multiplication tables are not only directly useful, but are also a foundation upon which other mathematical techniques are based – including multiplication of larger numbers, division, and fractions.

Required Skills Before Learning Times Tables:

Some very basic math skills are required before starting to learn times tables. These are:-

(1) Being able to count to over 100.

(2) Being able to add a single digit number to another single digit number.

(3) Being able to add a single digit number to a two digit number.

You should therefore make sure that your child is familiar and comfortable with these basics, before attempting to teach him multiplication tables!

The Structure of This Book:

This book is intended as a guide to help with parents and teachers teaching multiplication times tables to children, although of course it can be used by adult learners too.

- The introduction to this book, which you are reading now, is written for the parent, guardian or teacher.

- The rest of the book is written for the learner/child, and addresses the child directly as "you". I recommend that the parent or teacher read aloud from each chapter to the child – with a pen and paper handy, so any difficult points can be gone over immediately. During this process, take regular breaks to practice the times tables and techniques shown in the book – such as reciting the times table, counting up in the times table, writing the times table down, practicing the finger tricks, etc.

Key Points for Learning Times Tables:

Before we begin the main text of the book, there are a few things that I want to say:

(1) There are lots of different ways to either learn or otherwise calculate the numbers in each times table. It is perfectly okay for a child to use whatever technique suits them to get to the right answer. Even if a child initially uses a somewhat cumbersome technique to get the answer, with regular practice they will gradually improve. Eventually they will be able to immediately answer to any multiplication question – but that doesn't happen instantly.

(2) Some people seem to think that you must learn the times tables in numerical order, first the 1s, then the 2s, then 3s, then 4s, and so on. I am however more concerned about results than following arbitrary rules – and my experience is that there is a better and easier order to learn them in. This is the order that I use in this book. However, you don't have to follow my suggested order exactly – use the order that feels best for you or your child.

(3) When learning times tables, as well as practicing each new one as it is learned, don't forget to regularly go over the previous ones that have already been learned.

(4) As I have already mentioned, times tables are important in themselves, but they are also a foundation upon which other mathematical techniques are based. For this reason, I emphasize several things throughout this book:

- The relationship between multiplication and addition.
- It doesn't matter in which order you multiply two numbers together. For example 5×3 gives the same answer as 3×5.
- That multiplying any number by 0 gives a result of 0.
- That multiplying any number (except 0) by 1 gives a result of the original number.

(5) This book contains the times tables from 0×0 up to 12×12.

I have included the 0 times table, and the 0s in each other times table. This is because knowing how to multiply by 0, while relatively easy, is absolutely required once students move on to multiplying larger numbers.

Some schools and teachers only teach times tables up to 10×10. I however advocate learning up to 12×12. Learning the extra ones only requires a small amount of extra effort, but the 11 and 12 times tables are common enough both in daily life and in math exams, that this extra effort is easily worth it!

The Best Way to Learn:

As far as mastering each of the times tables is concerned, the best way to do this:

(1) Practice often and regularly – including practicing the tables that have previously been completed (so as not to forget).

(2) Practice each times tables multiple ways:

The more a student practices times tables, the better they will get. Additionally, the more ways a student practices, the better they will remember:

- Practice counting-up through the numbers in a particular times table, for example: "0, 2, 4, 6, 8…". Practice should include doing this verbally and by writing these numbers down – and yes, using fingers is allowed (even encouraged).
- Practice reciting each time tables table "0 times 2 is 0, 1 times 2 is 2, 2 times 2 is 4, 3 times 2 is 6, 4 times 2 is 8…" Again, this practice should be done both verbally and by writing them down.
- For some times tables there are "tricks" or "short-cuts" which can be used. For example, I will explain a finger trick for the 9 times table in this book. It is perfectly okay for students to use such tricks – and memorizing such tricks is often the best way to get started with a new times table.
- Practice randomly chosen questions from a particular times table.

(3) Do lots of activities involving times tables:

It's important to devote time to learning times tables, but you don't need to make it boring. Young children learn better when learning is fun! Additionally, you can make times tables part of your daily routine – put up a times table poster in your home, play a times table quiz when going on a car ride, etc.

Here some ideas that you can might consider:

- There are many games that you can play using times tables (including games that you can buy).

- There are many websites for practicing times tables, or for generating printable materials containing times tables.

- Get a multiplication tables poster, and place it in a prominent and highly visible location in your home.

I have placed some more information about these types of resources, including links, at http://www.suniltanna.com/multiplication

(4) Go slowly and allow students to work at their own pace:

There is no need to rush through the chapters. Spend as much time on each chapter as necessary before moving on: It is perfectly okay to spend several days working on a single chapter.

It is also okay for students to work-out the answers: <u>Eventually</u> the goal is to be able to answer times tables questions immediately, directly, and fluently, but until complete mastery is achieved (which can take weeks or even months), it is fine for students to work-out the answers using fingers, by counting, by repeated adding, or by using a "trick".

For example, when doing the 6 times table, a question might be "What is 9 times 6?"

- The student might know how to count-up in sixes ("0, 6, 12, 18..."), and unfold one finger at time until he reaches 9 times 6.
- The student might remember (perhaps with prompting) that 9 times 6 has the same answer as 6 times 9, and then if he knows the answer to the latter, be able to give the answer to the former.
- The student might remember (perhaps with prompting) that 9 times 6 has the same answer as 6 times 9, and then be able to use the 9s finger trick (explained later in this book) to work out 6 times 9.
- The student might know the answer to 8 times 6, and then add 6 more to this to get the answer to 9 times 6.

A student who is able to find the correct answer using any of these techniques will not only increase their confidence and sense of achievement, but will also soon improve their fluency in directly giving answers. In other words, applying brainpower to work-out answers seems to also help in improving memory!

How This Book is Organized:

Each chapter in the book after this one introduces a new topic:

- The next chapter introduces the concept of multiplication, and explains what multiplication is. This is a very important topic, so make sure your child really does understand before moving on.
- Most of the subsequent chapters introduce an additional times table – as well as some key facts about that times table.
- There are also a couple of extra chapters mixed into the times tables chapters, as well as a conclusion chapter.

Most chapters contain a combination of new facts and ideas, as well as reminders of other facts that have previously been covered.

- Important new facts and ideas are shown in **bold red**.

- Important facts and ideas that have been previously covered, but that are being recapped, are shown in **bold blue**.

Chapter 1: What is Multiplication?

In math, **the × symbol means multiply** two numbers together.

Additionally, **another word also meaning multiply is "times"**.

So, for example, 3 × 4, is read aloud as "3 multiplied by 4", or as "3 times 4".

But what does 3 × 4 mean?

Multiplication is simply the process of adding a number to itself again and again.

So 3 × 4 means 4 + 4 + 4 – that is 4 added to itself 3 times. Or, writing this in math symbols we would write:

$$3 \times 4 = 4 + 4 + 4$$

And if we worked out the answer to the addition question (it is 12), we would also have the answer to the multiplication question (which is also 12).

The answer to a multiplication question is called the product. So we could say "the product of 3 times 4 is 12", or "the product of 3 and 4 is 12".

One interesting fact about multiplication is that **it doesn't matter what order you do multiplication in.** Working out 4 × 3 will give you exactly the same answer as working out 3 × 4.

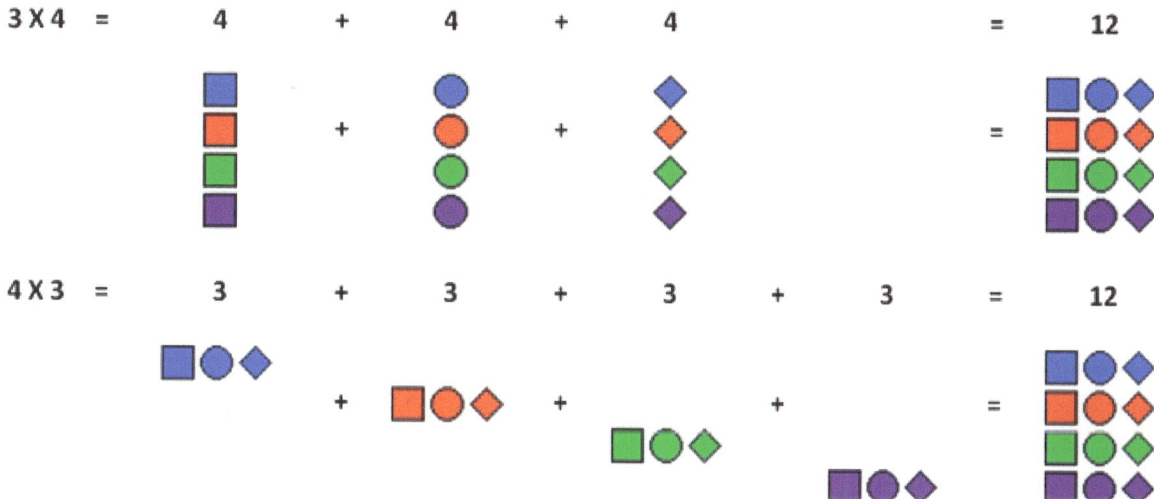

You can switch the numbers around like this for <u>any</u> multiplication question. It works for any pair of numbers that you might want to multiply together.

There is even a special phrase that mathematicians use to describe this fact, although you don't have to worry too much about remembering it. They say that "**multiplication is commutative**".

What is more important than remembering this phrase, is simply to remember that you can switch multiplication questions around in this way. This means that if you are asked a multiplication question and can't work out the answer one way round, simply try working it out the other way round!

This trick can also help when you start learning times tables. This is because whenever you begin a new times table, you will often be able to switch the questions around to turn them into questions that you already know the answers to.

Chapter 2: Learning the 0 Times Table

The 0 times table is so easy that some people don't even consider it a proper times table. In fact, you can learn it in the less than the time that it takes to write down your own name!

To learn the 0 times table, there is just one simple rule you need to learn: **Any number times 0 equals 0**.

So, for example:

- $1 \times 0 = 0$
- $3 \times 0 = 0$
- $7 \times 0 = 0$
- $12 \times 0 = 0$
- $37 \times 0 = 0$

Easy, isn't it?

- $0 \times 0 = 0$
- $1 \times 0 = 0$
- $2 \times 0 = 0$
- $3 \times 0 = 0$
- $4 \times 0 = 0$
- $5 \times 0 = 0$
- $6 \times 0 = 0$
- $7 \times 0 = 0$
- $8 \times 0 = 0$
- $9 \times 0 = 0$
- $10 \times 0 = 0$
- $11 \times 0 = 0$
- $12 \times 0 = 0$

Chapter 3: Learning the 1 Times Table

We've already learned that any number times 0 equals 0 so we can say:

- $1 \times 0 = 0$

We've also learned that it doesn't what order we do a multiplication problem in, so 1×0 will give the same answer as 0×1.

So we can begin the 1 times table with

- $1 \times 0 = 0 \times 1 = 0$

For the rest of the 1 times table is concerned, there's a simple rule: Any number (except 0) times 1 produces the original number.

So, for example:

- $5 \times 1 = 5$
- $7 \times 1 = 7$
- $12 \times 1 = 12$
- $327 \times 1 = 327$

So here is the 1 times table in full – as you can see, the answers in the 1 times table are just like counting:

- $0 \times 1 = 0$
- $1 \times 1 = 1$
- $2 \times 1 = 2$
- $3 \times 1 = 3$
- $4 \times 1 = 4$
- $5 \times 1 = 5$
- $6 \times 1 = 6$
- $7 \times 1 = 7$
- $8 \times 1 = 8$
- $9 \times 1 = 9$
- $10 \times 1 = 10$
- $11 \times 1 = 11$
- $12 \times 1 = 12$

Or, if we just prefer to write out the answers of 1 times table in order. It's exactly the same as counting:

- 0
- 1
- 2
- 3
- 4

- 5
- 6
- 7
- 8
- 9
- 10
- 11
- 12

Words for One

When something is made of one part (or can combined into one whole), the word used to describe it often begins with un- or uni-, or occasionally with mono-.

Here are some examples:

- A **unicycle** is a cycle with one wheel.
- To **unify** and to **unite** mean to combine several things/groups into one.
- A **monocle** is an eyeglass for one eye.
- A **monolith** is a monument or pillar made of one stone.
- A **monorail** is a type of train that moves on one rail

Chapter 4: Learning the 2 Times Table

The 2 times table is like counting-up by 2 at a time.

In other words, the 2 times table contains every second number:

- 0
- ...but not 1
- 2
- ...but not 3
- 4
- ...but not 5
- 6
- ...but not 7
- 8
- ...and so on

If we write the numbers in the 2 times table in order, the numbers are:

- 0
- 2
- 4
- 6
- 8
- 10
- 12
- 14
- 16
- 18
- 20
- 22
- 24

Additionally, you should already know the first two items on the 2 times table:

- 0×2 gives the same answer as 2×0. Any number times 0 equals 0. So therefore 0×2 equals 0.
- 1×2 gives the same answer as 2×1. Any number (except 0) times 1 equals the original number. So therefore 1×2 equals 2.

Here is the 2 times table in full:

- $0 \times 2 = 0$
- $1 \times 2 = 2$
- $2 \times 2 = 4$
- $3 \times 2 = 6$
- $4 \times 2 = 8$

- $5 \times 2 = 10$
- $6 \times 2 = 12$
- $7 \times 2 = 14$
- $8 \times 2 = 16$
- $9 \times 2 = 18$
- $10 \times 2 = 20$
- $11 \times 2 = 22$
- $12 \times 2 = 24$

Important Facts about the 2 Times Table

Here are some important facts to learn about the 2 times table:

- Numbers which are in the two times table are called even numbers. The first few even numbers are 0, 2, 4, 6, 8, 10, 12, 14, (and so on)
- Numbers which are not in the two times table are called odd numbers. The first few odd numbers are 1, 3, 5, 7, 9, 11, 13, 15, (and so on)

How to Tell if a Number is in the 2 Times Table

There is an easy way that you can look at **any** number and quickly tell whether or not it is in the 2 times table.

You do this by looking at the rightmost digit of the number:

- If the rightmost digit of the number is even, that is 0, 2, 4, 6, or 8, then the overall number is also an even number (in the 2 times table).
- If the rightmost digit of a number is odd, namely 1, 3, 5, 7, or 9, then the overall number is also an odd number (not in the 2 times table)

Here are some examples:

- 738 is an even number (in the 2 times table) – we know that, because its rightmost digit is 8, which is even.
- 543 is an odd number (**not** in the 2 times table) – we know that, because its rightmost digit is 3, which is odd.
- 618 is an even number (in the 2 times table) – we know that, because its rightmost digit is 8, which is even.
- 127 is an odd number (**not** in the 2 times table) – we know that, because its rightmost digit is 7, which is odd.
- 990 is an even number (in the 2 times table) – we know that, because its rightmost digit is 0, which is even.
- 1095 is an odd number (**not** in the 2 times table) – we know that, because its rightmost digit is 5, which is odd.

Words for Two

When we have a group of two things we sometimes call them a couple, a pair, or a duo.

When we multiply a number by 2, we double it.

Two times a number is said to be twice that number. For example, two times five, is said to be "twice five".

When something is made of two parts, or is commonly divided into two pieces, the word used to describe it often begins with bi- or bin-, or occasionally with di- or duo-.

Here are some examples:

- A **bicycle** is a cycle with two wheels.
- A **biped** is an animal that walks on two legs.
- A person who is **bilingual** speaks two languages.
- A sound we hear through both (two) ears is **binaural**.
- Something made of two pieces is **binary**.
- When we split something into at least two parts we **bisect** or **dissect** or **divide** it.
- A picture printed with two colors is a **duotone**.

Chapter 5: Learning the 10 Times Table

The 10 times table is very easy: **The 10 times table is just like counting, except that you put an extra zero to the right of the number** (except in the case of 0).

Thus, instead of counting 0, 1, 2, 3, 4, and so on... for the 10 times table we go 0, 10, 20, 30, 40, and so on...

If we write the numbers in the 10 times table in order, the numbers are:

- 0
- 10
- 20
- 30
- 40
- 50
- 60
- 70
- 80
- 90
- 100
- 110
- 120

Even without this trick, you should already know a few numbers on the 10 times table:

- 0×10 gives the same answer as 10×0. Any number times 0 equals 0. So therefore 0×10 must equal 0.
- 1×10 gives the same answer as 10×1. Any number (except 0) times 1 equals the original number. So therefore 1×10 equals 10.
- 2×10 gives the same answer as 10×2. From learning the 2 times table, you should know that 10×2 equals 20. So therefore 2×10 also equals 20.

Here is the 10 times table in full:

- $0 \times 10 = 0$
- $1 \times 10 = 10$
- $2 \times 10 = 20$
- $3 \times 10 = 30$
- $4 \times 10 = 40$
- $5 \times 10 = 50$
- $6 \times 10 = 60$
- $7 \times 10 = 70$
- $8 \times 10 = 80$
- $9 \times 10 = 90$
- $10 \times 10 = 100$
- $11 \times 10 = 110$

- $12 \times 10 = 120$

How to Tell if a Number is in the 10 Times Table

There is an easy way that you can look at **any** number and quickly tell whether or not it is in the 10 times table.

You do this by looking at the rightmost digit of the number:

- **If the rightmost digit of the number is 0, then the overall number is in the 10 times table.**
- **If the rightmost digit of a number is not 0, then the overall number is <u>not</u> in the 10 times table.**

Here are some examples:

- 730 is in the 10 times table – we know that because its rightmost digit is a 0.
- 543 is **<u>not</u>** in the 10 times table – we know that because its rightmost digit is not 0 (it's 3).
- 618 is **<u>not</u>** in the 10 times table – we know that, because its rightmost digit is not 0 (it's 8).
- 130 is in the 10 times table – we know that because its rightmost digit is a 0.
- 990 is in the 10 times table – we know that because its rightmost digit is a 0.
- 1095 is **<u>not</u>** in the 10 times table – we know that because its rightmost digit is not 0 (it's 5).

Words for Ten

When something is made of ten parts, or is commonly divided into ten pieces, the word used to describe it often begins with **dec -**.

Here are some examples:

- A **decade** is ten years.
- A **decagon** is a shape with ten sides and ten angles.
- A **decapod** is an animal with ten legs.
- A **decathlon** is sporting contest featuring ten different events.

Chapter 6: Learning the 5 Times Table

The 5 times table is a lot like the 10 times table, except in between each number in the 10 times table, we insert another number ending in 5:-

- 0 (in the 10 times table)
- 5
- 10 (in the 10 times table)
- 15
- 20 (in the 10 times table)
- 25
- 30 (in the 10 times table)
- 35
- ...and so on

If we write the numbers in the 5 times table in order, the numbers are:

- 0
- 5
- 10
- 15
- 20
- 25
- 30
- 35
- 40
- 45
- 50
- 55
- 60

Additionally, you already know several items on the 5 times table:

- 0×5 gives the same answer as 5×0. Any number times 0 equals 0. So therefore 0×5 equals 0.
- 1×5 gives the same answer as 5×1. Any number (except 0) times 1 equals the original number. So therefore 1×5 equals 5.
- 2×5 gives the same answer as 5×2. You already know 5×2 equals 10. So therefore 2×5 must also equal 10.
- 10×5 gives the same answer as 5×10. You already know 5×10 equals 50. So therefore 10×5 must also equal 50.

Here is the 5 times table in full:

- $0 \times 5 = 0$
- $1 \times 5 = 5$
- $2 \times 5 = 10$

- 3 × 5 = 15
- 4 × 5 = 20
- 5 × 5 = 25
- 6 × 5 = 30
- 7 × 5 = 35
- 8 × 5 = 40
- 9 × 5 = 45
- 10 × 5 = 50
- 11 × 5 = 55
- 12 × 5 = 60

How to Tell if a Number is in the 5 Times Table

There is an easy way that you can look at **any** number and quickly tell whether or not it is in the 5 times table.

You do this by looking at the rightmost digit of the number:

- **If the rightmost digit of the number is 0 or 5, then the overall number is in the 5 times table.**
- **If the rightmost digit of a number is neither 0 nor 5, then the overall number is <u>not</u> in the 5 times table.**

Here are some examples:

- 730 is in the 5 times table – we know that because its rightmost digit is a 0.
- 543 is **not** in the 5 times table – we know that because its rightmost digit is neither 0 nor 5 (it's 3).
- 618 is **not** in the 5 times table – we know that, because its rightmost digit is neither 0 nor 5 (it's 8).
- 130 is in the 5 times table – we know that because its rightmost digit is a 0.
- 990 is in the 5 times table – we know that because its rightmost digit is a 0.
- 1095 is in the 5 times table – we know that because its rightmost digit is a 5.

Words for Five

When something is made of five parts, or is commonly divided into five pieces, the word used to describe it often begins with pent -.

Here are some examples:

- A **pentagon** is a shape with five sides and five angles.
- A **pentathlon** is sporting contest featuring five different events.
- A **pentatonic** scale is a musical scale with five notes.

Chapter 7: Learning the 3 Times Table

The 3 times table is like counting-up by 3 at a time.

In other words, **the 3 times table contains every third number**:

- 0
- ...but not 1
- ...and not 2
- 3
- ...but not 4
- ...and not 5
- 6
- ...but not 7
- ...and not 8
- 9
- ...but not 10
- ...and not 11
- 12
- ...and so on

If we write the numbers in the 3 times table in order, the numbers are:

- **0**
- **3**
- **6**
- **9**
- **12**
- **15**
- **18**
- **21**
- **24**
- **27**
- **30**
- **33**
- **36**

Additionally, you already know several items on the 3 times table:

- 0×3 gives the same answer as 3×0. Any number times 0 equals 0. So therefore 0×3 equals 0.
- 1×3 gives the same answer as 3×1. Any number (except 0) times 1 equals the original number. So therefore 1×3 equals 3.
- 2×3 gives the same answer as 3×2. You already know 3×2 equals 6. So therefore 2×3 must also equal 6.

- 5 × 3 gives the same answer as 3 × 5. You already know 3 × 5 equals 15. So therefore 5 × 3 must also equal 15.
- 10 × 3 gives the same answer as 3 × 10. You already know 3 × 10 equals 30. So therefore 10 × 3 must also equal 30.

Here is the 3 times table in full:

- 0 × 3 = 0
- 1 × 3 = 3
- 2 × 3 = 6
- 3 × 3 = 9
- 4 × 3 = 12
- 5 × 3 = 15
- 6 × 3 = 18
- 7 × 3 = 21
- 8 × 3 = 24
- 9 × 3 = 27
- 10 × 3 = 30
- 11 × 3 = 33
- 12 × 3 = 36

How to Tell if a Number is in the 3 Times Table

There is an easy way that you can look at **any** number and quickly tell whether or not it is in the 3 times table.

You do this by adding up the digits that make up the number.

- If the digits add up to a number which is in the 3 times table, then the original number is also in the 3 times table.
- If the digits do not add up to a number which is in the 3 times table, then the original number is also not in the 3 times table.

Here are some examples:

- 738 – adding the digits (7 + 3 + 8) gives 18. As 18 is in the 3 times table, we know that 738 is in the 3 times table.
- 543 – adding the digits (5 + 4 + 3) gives 12. As 12 is in the 3 times table, we know that 543 is in the 3 times table.
- 127 – adding the digits (1 + 2 + 7) gives 10. As 10 is **not** in the 3 times table, we know that 127 is **not** in the 3 times table.
- 990 – adding the digits (9 + 9 + 0) gives 18. As 18 is in the 3 times table, we know that 990 is in the 3 times table.
- 1095 – adding the digits (1 + 0 + 9 + 5) gives 15. As 15 is in the 3 times table, we know that 1095 is in the 3 times table.

- 638 – adding the digits (6 + 3 + 8) gives 17. As 17 is **not** in the 3 times table, we know that 638 is **not** in the 3 times table.

Words for Three

A group of three is sometimes called a trio.

When we multiply a number by 3, we triple it.

Three times a number is said to be thrice that number. For example, three times five, is said to be "thrice five".

When something is made of three parts, or is commonly divided into three pieces, the word used to describe it often begins with tri-.

Here are some examples:

- A **triangle** is a shape with three sides and three angles.
- A **triathlon** is a sporting contest featuring three different events.
- A **tricycle** is a cycle with three wheels.
- A **tripod** is a stool or table with three legs.
- Three children born at the same time are **triplets**.

Chapter 8: Learning the 11 Times Table

Like all the other times tables, the 11 times table begins with 0 (because 0 × 11 equals 0).

Most of the numbers in the 11 times table are easy to remember: simply repeat each digit twice – 11, 22, 33, 44, 55, 66, 77, 88, and 99.

However you do need to remember the next three numbers: 110, 121, and 132.

If we write the numbers in the 11 times table in order, the numbers are:

- 0
- 11
- 22
- 33
- 44
- 55
- 66
- 77
- 88
- 99
- 110
- 121
- 132

Even without this trick, you should already know a few numbers on the 11 times table:

- 0 × 11 gives the same answer as 11 × 0. Any number times 0 equals 0. So therefore 0 × 11 must equal 0.
- 1 × 11 gives the same answer as 11 × 1. Any number (except 0) times 1 equals the original number. So therefore 1 × 11 equals 11.
- 2 × 11 gives the same answer as 11 × 2. From learning the 2 times table, you should know that 11 × 2 equals 22. So therefore 2 × 11 also equals 22.
- 3 × 11 gives the same answer as 11 × 3. From learning the 3 times table, you should know that 11 × 3 equals 33. So therefore 3 × 11 also equals 33.
- 5 × 11 gives the same answer as 11 × 5. From learning the 5 times table, you should know that 11 × 5 equals 55. So therefore 5 × 11 also equals 55.
- 10 × 11 gives the same answer as 11 × 10. From learning the 10 times table, you should know that 11 × 10 equals 110. So therefore 10 × 11 also equals 110.

Here is the 11 times table in full:

- 0 × 11 = 0
- 1 × 11 = 11
- 2 × 11 = 22
- 3 × 11 = 33
- 4 × 11 = 44

- $5 \times 11 = 55$
- $6 \times 11 = 66$
- $7 \times 11 = 77$
- $8 \times 11 = 88$
- $9 \times 11 = 99$
- $10 \times 11 = 110$
- $11 \times 11 = 121$
- $12 \times 11 = 132$

Chapter 9: Learning the 9 Times Table

Like all the other times tables, the 9 times table begins with 0 (because 0 × 9 equals 0).

There are three different **tricks that you can use to help remember the rest of 9 times table**. (I prefer the finger trick - and would recommend starting with that - but it is okay to use whichever trick suits you best).

The Finger Trick:

This is an easy trick to find most of the numbers in the 9 times table.

Put both hands in front of you, palms down, with fingers spread apart, like this:

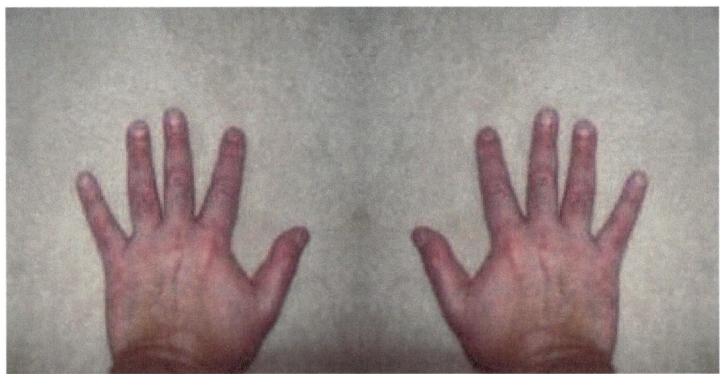

Now choose the number that you wish to multiply 9 by. Counting from the left, find the corresponding finger, and bend that finger down. For example, if you wished to work out 4 × 9, you would find the 4th finger from the left, and bend it like this:

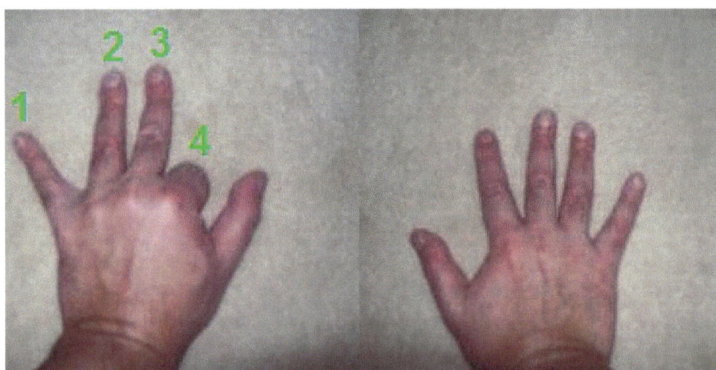

The number of fingers to the left of the bent finger gives the tens column. The number of fingers to the right of the bent finger gives the units column.

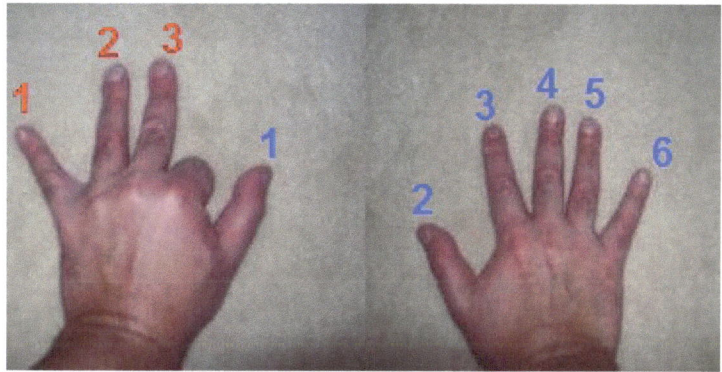

So, in this example, the digits in the answer are 3 and 6, and so answer is 36.

Let's try doing another question: What is 7 × 9 ?

First we find our 7th finger and bend it down:

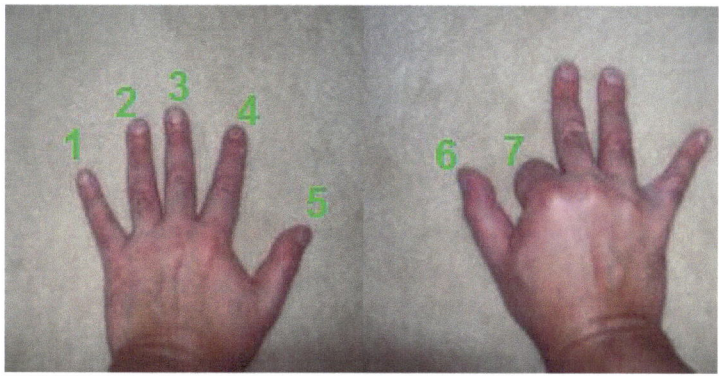

Then we count the fingers before and after the bent finger.

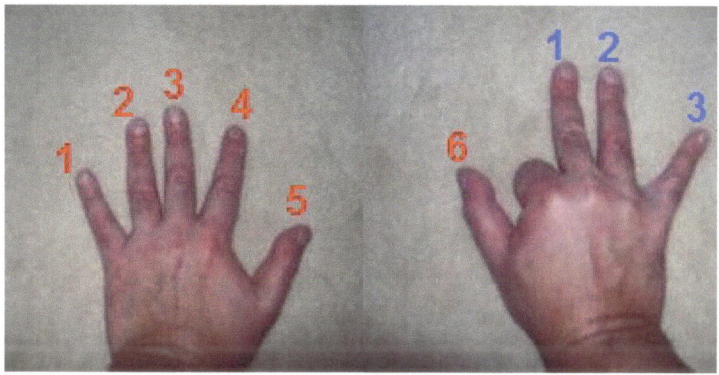

The digits in the answer are 6 and 3. So the answer equals 63.

Please Note:

(1) If you use this trick, you will still have to remember 0 × 9 equals 0, as well as the answers to 11 × 9 (99), and 12 × 9 (108).

(2) Many people find this such a handy trick that they use it to work it 9 × on the other times tables. For example, if you want to know the answer to 9 × 7, you can simply work out 7 × 9 using this finger trick!

The Digits Trick:

This is another trick that you can also use to help with most of the 9 times table. This trick is generally most helpful if you can take a moment to write down the numbers in the 9 times table.

Simply remember that the digit in the tens column increases by 1 each time, and the number in the units column decreases by 1 each time:

- Tens = 0, Units = 9, or in combination 0 9 – or simply 9
- Tens = 1, Units = 8, or in combination 1 8 – or simply 18
- Tens = 2, Units = 7, or in combination 2 7 – or simply 27
- Tens = 3, Units = 6, or in combination 3 6 – or simply 36
- Tens = 4, Units = 5, or in combination 4 5 – or simply 45
- Tens = 5, Units = 4, or in combination 5 4 – or simply 54
- Tens = 6, Units = 3, or in combination 6 3 – or simply 63
- Tens = 7, Units = 2, or in combination 7 2 – or simply 72
- Tens = 8, Units = 1, or in combination 8 1 – or simply 81
- Tens = 9, Units = 0, or in combination 9 0 – or simply 90

Please Note:

(1) You may find it helpful to remember that in each case, the pair of digits always adds up to 9. For example, 63 is in the 9 times table, and 6 + 3 equals 9.

(2) If you use this trick, you will still have to remember 0 × 9 equals 0, as well as the answers to 11 × 9 (99), and 12 × 9 (108).

Times 10 then Take Away Trick:

Most people find the 10 times table to be very easy. If you are good at taking away, you can use the 10 times table to help with the 9s: To multiply a number by 9, first multiply by 10, then take away the number.

Here are some examples:

- Working out 7 × 9: First work out that 7 × 10 is 70, then take away 7. 70 take away 7 is 63, so 7 × 9 is also 63.
- Working out 5 × 9: First work out that 5 × 10 is 50, then take away 5. 50 take away 5 is 45, so 5 × 9 is also 45.
- Working out 9 × 9: First work out that 9 × 10 is 90, then take away 9. 90 take away 9 is 81, so 9 × 9 is also 81.

The 9 Times Table in Full

Even without using these tricks, you should already know quite a few numbers on the 9 times table:

- 0×9 gives the same answer as 9×0. Any number times 0 equals 0. So therefore 0×9 must equal 0.
- 1×9 gives the same answer as 9×1. Any number (except 0) times 1 equals the original number. So therefore 1×9 equals 9.
- 2×9 gives the same answer as 9×2. From learning the 2 times table, you should know that 9×2 equals 18. So therefore 2×9 also equals 18.
- 3×9 gives the same answer as 9×3. From learning the 3 times table, you should know that 9×3 equals 27. So therefore 3×9 also equals 27.
- 5×9 gives the same answer as 9×5. From learning the 5 times table, you should know that 9×5 equals 45. So therefore 5×9 also equals 45.
- 10×9 gives the same answer as 9×10. From learning the 10 times table, you should know that 9×10 equals 90. So therefore 10×9 also equals 90.
- 11×9 gives the same answer as 9×11. From learning the 11 times table, you should know that 9×11 equals 99. So therefore 11×9 also equals 99.

Here is the 9 times table in full:

- $0 \times 9 = 0$
- $1 \times 9 = 9$
- $2 \times 9 = 18$
- $3 \times 9 = 27$
- $4 \times 9 = 36$
- $5 \times 9 = 45$
- $6 \times 9 = 54$
- $7 \times 9 = 63$
- $8 \times 9 = 72$
- $9 \times 9 = 81$
- $10 \times 9 = 90$
- $11 \times 9 = 99$
- $12 \times 9 = 108$

If we list the numbers in the 9 times table in order, the numbers are:

- 0
- 9
- 18
- 27
- 36
- 45
- 54
- 63
- 72
- 81
- 90
- 99

How to Tell if a Number is in the 9 Times Table

There is an easy way that you can look at **any** number and quickly tell whether or not it is in the 9 times table.

You do this by adding up the digits that make up the number.

- If the digits add up to a number which is in the 9 times table, then the original number is also in the 9 times table.
- If the digits do not add up to a number which is in the 9 times table, then the original number is also not in the 9 times table.

Here are some examples:

- 738 – adding the digits (7 + 3 + 8) gives 18. As 18 is in the 9 times table, we know that 738 is in the 9 times table.
- 543 – adding the digits (5 + 4 + 3) gives 12. As 12 is **not** in the 9 times table, we know that 543 is **not** in the 9 times table.
- 127 – adding the digits (1 + 2 + 7) gives 10. As 10 is **not** in the 9 times table, we know that 127 is **not** in the 9 times table.
- 990 – adding the digits (9 + 9 + 0) gives 18. As 18 is in the 9 times table, we know that 990 is in the 9 times table.
- 1095 – adding the digits (1 + 0 + 9 + 5) gives 15. As 15 is **not** in the 9 times table, we know that 1095 is **not** in the 9 times table.

Words for Nine

When something is made of nine parts, or is commonly divided into nine pieces, the word used to describe it often begins with **non-** or sometimes **nov-**.

Here are some examples:

- A **nonagon** is a shape with nine sides and nine angles.
- A **nonet** is a group of nine musicians.
- A **novena** is a set of religious prayers that last for nine days.
- A **novennial** event is one that occurs every nine years.

Chapter 10: Learning the 4 Times Table

If you are able to count-up in 2s, then you already nearly know the 4 times table.

The 4 times table contains every second number in the 2 times table:

- 0
- ...but not 2
- 4
- ...but not 6
- 8
- ...but not 10
- 12
- ...but not 14
- 16
- ...and so on

If we write the numbers in the 4 times table in order, the numbers are:

- 0
- 4
- 8
- 12
- 16
- 20
- 24
- 28
- 32
- 36
- 40
- 44
- 48

Additionally, you should already know several of the items on the 4 times table:

- 0×4 gives the same answer as 4×0. Any number times 0 equals 0. So therefore 0×4 must equal 0.
- 1×4 gives the same answer as 4×1. Any number (except 0) times 1 equals the original number. So therefore 1×4 equals 4.
- 2×4 gives the same answer as 4×2. From learning the 2 times table, you should know that 4×2 equals 8. So therefore 2×4 also equals 8.
- 3×4 gives the same answer as 4×3. From learning the 3 times table, you should know that 4×3 equals 12. So therefore 3×4 also equals 12.
- 5×4 gives the same answer as 4×5. From learning the 5 times table, you should know that 4×5 equals 20. So therefore 5×4 also equals 20.

- 9 × 4 gives the same answer as 4 × 9. From learning the 9 times table, you should know that 4 × 9 equals 36. So therefore 9 × 4 also equals 36.
- 10 × 4 gives the same answer as 4 × 10. From learning the 10 times table, you should know that 4 × 10 equals 40. So therefore 10 × 4 also equals 40.
- 11 × 4 gives the same answer as 4 × 11. From learning the 11 times table, you should know that 4 × 11 equals 44. So therefore 11 × 4 also equals 44.

Here is the 4 times table in full:

- 0 × 4 = 0
- 1 × 4 = 4
- 2 × 4 = 8
- 3 × 4 = 12
- 4 × 4 = 16
- 5 × 4 = 20
- 6 × 4 = 24
- 7 × 4 = 28
- 8 × 4 = 32
- 9 × 4 = 36
- 10 × 4 = 40
- 11 × 4 = 44
- 12 × 4 = 48

How to Tell if a Number is in the 4 Times Table

There are methods which use on **any** number to tell whether or not it is in the 4 times table. These methods are not quite as easy as some of the other tables (I would **not** worry about learning these methods and you may wish to skip over this section).

Anyway, here is one way in which you can tell if a number is in the 4 times table:

Step 1: **If the number is odd (not on the two times table), then the number is _not_ on the 4 times table**.

Step 2 (optional but a time-saver): Look at the two rightmost digits:
- **If the two rightmost digits of a number correspond to a number on the 4 times table, then the overall number is also on the 4 times table.**
- **If the two rightmost digits of a number correspond to a number _not_ on the 4 times table, then the overall number is also _not_ on the 4 times table.**
- If you are still unsure, use step 3 to decide.

Step 3: Look at the two rightmost digits:
- **If the tens column is odd (1, 3, 5, 7 or 9) – the overall number is in the 4 times table only if the units column contains 2 or 6.**

Here are some examples of using this method:

- 736 is an even number, so we need to look its rightmost two digits. These are 36. We know 36 is in the 4 times table, so therefore 736 must also in the 4 times table.
- 543 is an odd number, so it can **not** be in the 4 times table.
- 618 is an even number, so we need to look its rightmost two digits. These are 18. We know 18 is **not** in the 4 times table, so therefore 618 is also **not** in the 4 times table.
- 127 is an odd number, so it can **not** be in the 4 times table.
- 984 is an even number, so we need to look its rightmost two digits. These are 84. If we knew the 4 times table all the way up to 84 and beyond, we would immediately be able to tell. If we aren't sure, we instead need to look at the tens column (even), and note that the units column is 4 – which according to the rules given earlier, means 984 is in the 4 times table.
- 1092 is an even number, so we need to look its rightmost two digits. These are 92. If we knew the 4 times table all the way up to 92 and beyond, we would immediately be able to tell. If we aren't sure, we instead need to look at the tens column (odd), and note that the units column is 2 – which according to the rules given earlier, means 1092 is in the 4 times table.
- 1374 is an even number, so we need to look its rightmost two digits. These are 74. If we knew the 4 times table all the way up to 74 and beyond, we would immediately be able to tell. If we aren't sure, we instead need to look at the tens column (odd), and note that the units column is 4 – which according to the rules given earlier, means 1374 is **not** in the 4 times table.

There is also an alternative method you can use, which involves halving. If you are good at halving numbers, you could use this method:

Step 1: **If the number is odd (not on the two times table), then the number is <u>not</u> on the 4 times table**.

Step 2 (optional but a time-saver): Look at the two rightmost digits:
- **If the two rightmost digits of a number correspond to a number on the 4 times table, then the overall number is also on the 4 times table.**
- **If the two rightmost digits of a number correspond to a number not on the 4 times table, then the overall number is also <u>not</u> on the 4 times table.**
- If you are still unsure, use step 3 to decide.

Step 3: Look at the two rightmost digits, and halve them.
- **If half of the number given by the two rightmost digits is even (ending with 0, 2, 4, 6 or 8) – the overall number is in the 4 times table.**
- **If half of the number given by the two rightmost digits is odd (ending with 1, 3, 5, 7 or 9) – the overall number is in the 4 times table.**

Here are some examples of using this method:

- 736 is an even number, so we need to look its rightmost two digits. These are 36. We know 36 is in the 4 times table, so therefore 736 is in the 4 times table.
- 543 is an odd number, so it can **not** be in the 4 times table.
- 618 is an even number, so we need to look its rightmost two digits. These are 18. We know 18 is **not** in the 4 times table, so therefore 618 is also **not** in the 4 times table.
- 127 is an odd number, so it can **not** be in the 4 times table.
- 984 is an even number, so we need to look its rightmost two digits. These are 84. If we knew the 4 times table all the way up to 84 and beyond, we would immediately be able to tell. If we aren't sure, we can work out that half of 84 is 42. Since 42 is even, that means that both 84 and 984 are in the 4 times table.
- 1092 is an even number, so we need to look its rightmost two digits. These are 92. If we knew the 4 times table all the way up to 92 and beyond, we would immediately be able to tell. If we aren't sure, we can work out that half of 92 is 46. Since 46 is even, that means that both 92 and 1092 are in the 4 times table.
- 1374 is an even number, so we need to look its rightmost two digits. These are 74. If we knew the 4 times table all the way up to 74 and beyond, we would immediately be able to tell. If we aren't sure, we can work out that half of 74 is 37. Since 37 is odd, that means that both 74 and 1374 are **not** in the 4 times table.

Words for Four

When we multiply a number by 4, we quadruple it.

Four times a number is said to be quadruple that number. For example, four times five, is said to be "quadruple five".

When something is made of four parts, or is commonly divided into four pieces, the word used to describe it often begins with quad-, quart-, or occasionally with tetra-.

Here are some examples:

- A **quadrilateral** is a shape with four sides and five angles.
- A **quadruped** is an animal that has four feet.
- Four children born at the same time are **quadruplets**.
- A **quarter** is a piece of something that has been split into four pieces.
- A **quartet** is a group of four musicians.
- A **tetrapod** is an animal with four limbs.
- A **tetrahedron** is a solid shape with four faces.

Chapter 11: Another Finger Trick to Help with the 6s, 7s, 8s, 9s, and 10s

In the last few chapters of this book, we will learn the remaining times tables.

Before that, I want to show you a handy finger trick that you can use to help you with some of the higher numbers in the 6, 7, 8, 9, and 10 times tables. It's especially helpful with the 6s, 7s, and 8s, which many people find difficult to learn.

Begin by putting your hands in front of you like this:

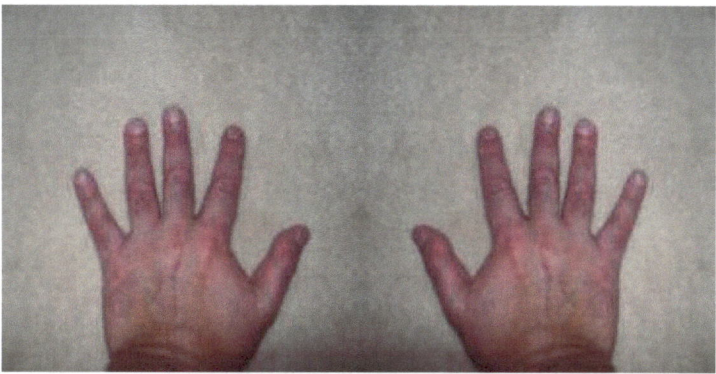

Now count the numbers from 6 upwards starting at each thumb:

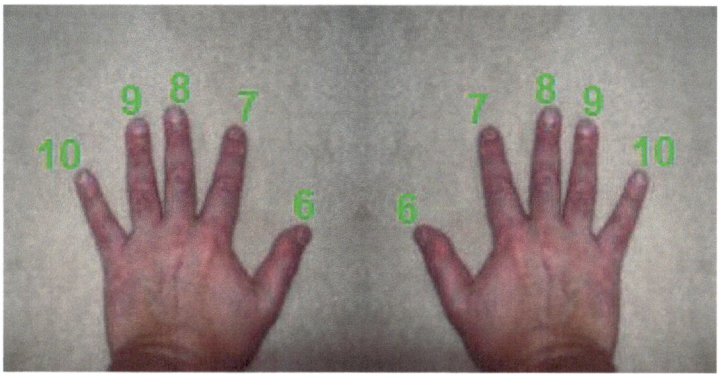

Now turn your hands inward and touch together the two fingers corresponding to the numbers that you wish to multiply.

For example, if you wanted to calculate 8 × 8, you would touch your left 8 finger to your right 8 finger, as shown in this picture:

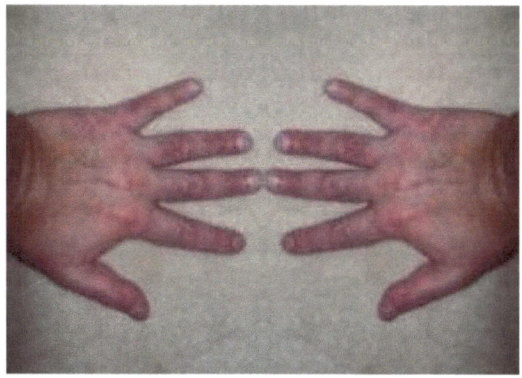

The answer comes from following three simple steps:

(1) Count the two touching fingers as well as all the fingers nearer than them. This gives the number of tens in the answer.

So, in this example, the tens column of the answer would be 6:

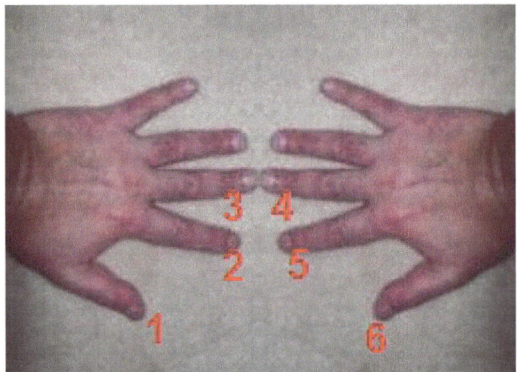

(2) Count the number of fingers on each hand that are farther away than the two touching fingers, and multiply them together to get the number in the units column.

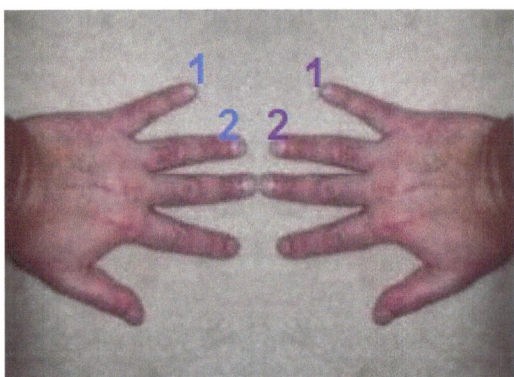

In this example, the farther away fingers on the left hand is 2, and on the right hand is also 2, giving a units value of 2 × 2, which is 4.

(3) Combine the two parts.

In this example, the tens column contains 6, the units column contains 4, and the overall answer (to 8 × 8) is therefore 64.

Chapter 12: Learning the 6 Times Table

If you are able to count-up in 3s, then you already nearly know the 6 times table.

The 6 times table contains every second number in the 3 times table:

- 0
- ...but not 3
- 6
- ...but not 9
- 12
- ...but not 15
- 18
- ...but not 21
- 24
- ...and so on

If we write the numbers in the 6 times table in order, the numbers are:

- 0
- 6
- 12
- 18
- 24
- 30
- 36
- 42
- 48
- 54
- 60
- 66
- 72

You should already know several of the items on the 6 times table:

- 0×6 gives the same answer as 6×0. Any number times 0 equals 0. So therefore 0×6 must equal 0.
- 1×6 gives the same answer as 6×1. Any number (except 0) times 1 equals the original number. So therefore 1×6 equals 6.
- 2×6 gives the same answer as 6×2. From learning the 2 times table, you should know that 6×2 equals 12. So therefore 2×6 also equals 12.
- 3×6 gives the same answer as 6×3. From learning the 3 times table, you should know that 6×3 equals 18. So therefore 3×6 also equals 18.
- 4×6 gives the same answer as 6×4. From learning the 4 times table, you should know that 6×4 equals 24. So therefore 4×6 also equals 24.

- 5 × 6 gives the same answer as 6 × 5. From learning the 5 times table, you should know that 6 × 5 equals 30. So therefore 5 × 6 also equals 30.
- 9 × 6 gives the same answer as 6 × 9. From learning the 9 times table, you should know that 6 × 9 equals 54. So therefore 9 × 6 also equals 54.
- 10 × 6 gives the same answer as 6 × 10. From learning the 10 times table, you should know that 6 × 10 equals 60. So therefore 10 × 6 also equals 60.
- 11 × 6 gives the same answer as 6 × 11. From learning the 11 times table, you should know that 6 × 11 equals 66. So therefore 11 × 6 also equals 66.

Here is the 6 times table in full:

- 0 × 6 = 0
- 1 × 6 = 6
- 2 × 6 = 12
- 3 × 6 = 18
- 4 × 6 = 24
- 5 × 6 = 30
- 6 × 6 = 36
- 7 × 6 = 42
- 8 × 6 = 48
- 9 × 6 = 54
- 10 × 6 = 60
- 11 × 6 = 66
- 12 × 6 = 72

Tricks to Help with the 6 Times Table

There is a little trick that you can use to help you remember the 6s: when multiplying an even number by 6, the rightmost digit is the same in both the question and answer: 2 × 6 = 12... 4 × 6 = 24... 6 × 6 = 36... 8 × 6 = 18... 10 × 6 = 60... 12 × 6 = 72.

Additionally, don't forget that you can use the 6, 7, 8, 9, 10 finger trick that we described in the last chapter. This trick can help you with several of the numbers on the 6 times table.

How to Tell if a Number is in the 6 Times Table

There is a relatively easy way that you can look at any number and quickly tell whether or not it is in the 6 times table.

Here is how you do it:

- If the number is both even (rightmost digit is 0, 2, 4, 6 or 8) and in the 3 times table (the digits add up to a number which is in the 3 times table), then the number is also in the 6 times table.

Here are some examples:

- 738 – is both even and in the 3 times table (adding the digits 7 + 3 + 8 gives 18, and 18 is in the 3 times table), so 738 is in the 6 times table.
- 543 – is odd, so is **not** in the 6 times table.
- 127 – is odd, so is **not** in the 6 times table.
- 990 – is both even and in the 3 times table (adding the digits 9 + 9 + 0 gives 18, and 18 is in the 3 times table). So 990 is in the 6 times table.
- 127 – is odd, so is **not** in the 6 times table.
- 1095 – is odd, so is **not** in the 6 times table.
- 638 – although even, 638 is **not** in the 3 times table (adding the digits 6 + 3 + 8 gives 17, and 17 is **not** in the 3 times table). So 638 is **not** in the 6 times table.

Words for Six

When something is made of six parts, or is commonly divided into six pieces, the word used to describe it often begins with **hex-**, or **sex-**.

Here are some examples:

- A **hexagon** is a shape with six sides and six angles.
- A **sextet** is a group of six musicians.

Chapter 13: Learning the 8 Times Table

If you are able to count-up in 4s, then you already know much of the 8 times table.

The 8 times table contains every second number in the 4 times table:

- 0
- ...but not 4
- 8
- ...but not 12
- 16
- ...but not 20
- 24
- ...but not 28
- 32
- ...and so on

If we write the numbers in the 8 times table in order, the numbers are:

- 0
- 8
- 16
- 24
- 32
- 40
- 48
- 56
- 64
- 72
- 80
- 88
- 96

You should already know several of the items on the 8 times table:

- 0 × 8 gives the same answer as 8 × 0. Any number times 0 equals 0. So therefore 0 × 8 must equal 0.
- 1 × 8 gives the same answer as 8 × 1. Any number (except 0) times 1 equals the original number. So therefore 1 × 8 equals 8.
- 2 × 8 gives the same answer as 8 × 2. From learning the 2 times table, you should know that 8 × 2 equals 16. So therefore 2 × 8 also equals 16.
- 3 × 8 gives the same answer as 8 × 3. From learning the 3 times table, you should know that 8 × 3 equals 24. So therefore 3 × 8 also equals 24.
- 4 × 8 gives the same answer as 8 × 4. From learning the 4 times table, you should know that 8 × 4 equals 32. So therefore 4 × 8 also equals 32.

- 5 × 8 gives the same answer as 8 × 5. From learning the 5 times table, you should know that 8 × 5 equals 40. So therefore 5 × 8 also equals 40.
- 6 × 8 gives the same answer as 8 × 6. From learning the 6 times table, you should know that 8 × 6 equals 48. So therefore 6 × 8 also equals 48.
- 9 × 8 gives the same answer as 8 × 9. From learning the 9 times table, you should know that 8 × 9 equals 72. So therefore 9 × 8 also equals 72.
- 10 × 8 gives the same answer as 8 × 10. From learning the 10 times table, you should know that 8 × 10 equals 80. So therefore 10 × 8 also equals 80.
- 11 × 8 gives the same answer as 8 × 11. From learning the 11 times table, you should know that 8 × 11 equals 88. So therefore 11 × 8 also equals 88.

Also don't forget that you can use the 6, 7, 8, 9, 10 finger trick that we described a couple of chapters ago. This trick can help you with several of the numbers on the 8 times table.

Here is the 8 times table in full:

- 0 × 8 = 0
- 1 × 8 = 8
- 2 × 8 = 16
- 3 × 8 = 24
- 4 × 8 = 32
- 5 × 8 = 40
- 6 × 8 = 48
- 7 × 8 = 56
- 8 × 8 = 64
- 9 × 8 = 72
- 10 × 8 = 80
- 11 × 8 = 88
- 12 × 8 = 96

Words for Eight

When something is made of eight parts, or Is commonly divided into eight pieces, the word used to describe it often begins with oct-.

Here are some examples:

- An **octagon** is a shape with eight sides and eight angles.
- An **octopus** is an animal with eight tentacles.

Chapter 14: Learning the 7 Times Table

The 7 times table is one of the trickiest to learn – or it would be if you had to learn it from scratch.

Fortunately however, by now, you should already know most of the 7 times table!

- 0 × 7 gives the same answer as 7 × 0. Any number times 0 equals 0. So therefore 0 × 7 must equal 0.
- 1 × 7 gives the same answer as 7 × 1. Any number (except 0) times 1 equals the original number. So therefore 1 × 7 equals 7.
- 2 × 7 gives the same answer as 7 × 2. From learning the 2 times table, you should know that 7 × 2 equals 14. So therefore 2 × 7 also equals 14.
- 3 × 7 gives the same answer as 7 × 3. From learning the 3 times table, you should know that 7 × 3 equals 21. So therefore 3 × 7 also equals 21.
- 4 × 7 gives the same answer as 7 × 4. From learning the 4 times table, you should know that 7 × 4 equals 28. So therefore 4 × 7 also equals 28.
- 5 × 7 gives the same answer as 7 × 5. From learning the 5 times table, you should know that 7 × 5 equals 35. So therefore 5 × 7 also equals 35.
- 6 × 7 gives the same answer as 7 × 6. From learning the 6 times table, you should know that 7 × 6 equals 42. So therefore 6 × 7 also equals 42.
- 8 × 7 gives the same answer as 7 × 8. From learning the 8 times table, you should know that 7 × 8 equals 56. So therefore 8 × 7 also equals 56.
- 9 × 7 gives the same answer as 7 × 9. From learning the 9 times table, you should know that 7 × 9 equals 63. So therefore 9 × 7 also equals 63.
- 10 × 7 gives the same answer as 7 × 10. From learning the 10 times table, you should know that 7 × 10 equals 70. So therefore 10 × 7 also equals 70.
- 11 × 7 gives the same answer as 7 × 11. From learning the 11 times table, you should know that 7 × 11 equals 77. So therefore 11 × 7 also equals 77.

Here is the 7 times table in full:

- 0 × 7 = 0
- 1 × 7 = 7
- 2 × 7 = 14
- 3 × 7 = 21
- 4 × 7 = 28
- 5 × 7 = 35
- 6 × 7 = 42
- 7 × 7 = 49
- 8 × 7 = 56
- 9 × 7 = 63
- 10 × 7 = 70
- 11 × 7 = 77
- 12 × 7 = 84

Additionally, don't forget that you can use the 6, 7, 8, 9, 10 finger trick that we described a few chapters ago. This trick can help you with several of the numbers on the 7 times table.

If we write the numbers in the 7 times table in order, the numbers are:

- 0
- 7
- 14
- 21
- 28
- 35
- 42
- 49
- 56
- 63
- 70
- 77
- 84

Words for Seven

When something is made of seven parts, or is commonly divided into seven pieces, the word used to describe it often begins with **hept-** or **sept-**.

Here are some examples:

- A **heptagon** is a shape with seven sides and seven angles.
- A **heptathlon** is sporting contest featuring seven different events.
- A **septennial** is a period of seven years.

Chapter 15: Learning the 12 Times Table

Some people think the 12 times table is hard, but there are quite a few tricks that you can use to help you. Additionally, if you've already learned all the other times tables already, you will discover that you have already learned nearly all of the 12 times table!

The first trick is that if you are able to count-up in 6s, then you already know much of the 12 times table.

The 12 times table contains every second number in the 6 times table:

- 0
- ...but not 6
- 12
- ...but not 18
- 24
- ...but not 30
- 36
- ...but not 42
- 48
- ...and so on

If we write the numbers in the 12 times table in order, the numbers are:

- **0**
- **12**
- **24**
- **36**
- **48**
- **60**
- **72**
- **84**
- **96**
- **108**
- **120**
- **132**
- **144**

As mentioned previously, if you've already learned the other times tables, you in fact will have already learned most of the 12 times table:

- 0 × 12 gives the same answer as 12 × 0. Any number times 0 equals 0. So therefore 0 × 12 must equal 0.
- 1 × 12 gives the same answer as 12 × 1. Any number (except 0) times 1 equals the original number. So therefore 1 × 12 equals 12.
- 2 × 12 gives the same answer as 12 × 2. From learning the 2 times table, you should know that 12 × 2 equals 24. So therefore 2 × 12 also equals 24.

- 3 × 12 gives the same answer as 12 × 3. From learning the 3 times table, you should know that 12 × 3 equals 36. So therefore 3 × 12 also equals 36.
- 4 × 12 gives the same answer as 12 × 4. From learning the 4 times table, you should know that 12 × 4 equals 48. So therefore 4 × 12 also equals 48.
- 5 × 12 gives the same answer as 12 × 5. From learning the 5 times table, you should know that 12 × 5 equals 60. So therefore 5 × 12 also equals 60.
- 6 × 12 gives the same answer as 12 × 6. From learning the 6 times table, you should know that 12 × 6 equals 72. So therefore 6 × 12 also equals 72.
- 7 × 12 gives the same answer as 12 × 7. From learning the 7 times table, you should know that 12 × 7 equals 84. So therefore 7 × 12 also equals 84.
- 8 × 12 gives the same answer as 12 × 8. From learning the 8 times table, you should know that 12 × 8 equals 96. So therefore 8 × 12 also equals 96.
- 9 × 12 gives the same answer as 12 × 9. From learning the 9 times table, you should know that 12 × 9 equals 108. So therefore 9 × 12 also equals 108.
- 10 × 12 gives the same answer as 12 × 10. From learning the 10 times table, you should know that 12 × 10 equals 120. So therefore 10 × 12 also equals 120.
- 11 × 12 gives the same answer as 12 × 11. From learning the 11 times table, you should know that 12 × 11 equals 132. So therefore 11 × 12 also equals 132.

Here is the 12 times table in full:

- 0 × 12 = 0
- 1 × 12 = 12
- 2 × 12 = 24
- 3 × 12 = 36
- 4 × 12 = 48
- 5 × 12 = 60
- 6 × 12 = 72
- 7 × 12 = 84
- 8 × 12 = 96
- 9 × 12 = 108
- 10 × 12 = 120
- 11 × 12 = 132
- 12 × 12 = 144

Tricks to Help with the 6 Times Table

If you struggle with the 12 times table, here are two tricks that might help to work out what a number times 12 is:

(1) To multiply a number by 12: First multiply the original number by 10, then multiply the original number by 2, and finally add the two results together.

For example, if you wanted to work out 7 × 12: You would work out that 7 × 10 is 70, then work out that 7 × 2 is 14. Then you would add 70 and 14 to get 84 as your overall answer.

Or (2) To multiply a number by 12: First work out what the number is multiplied by 6, then double that value to get the overall answer.

For example, if you wanted to work out 7 × 12: You would work out that 7 × 6 is 42. You would then double 42 to get your overall answer of 84.

Words for Twelve

The word dozen is used to refer to a group of 12 items (for example: "a dozen eggs").

Sometimes people may count in multiples of dozens. For example:

- **Two dozen** would be 2 × 12 (which is 24).
- **Three dozen** would be 3 × 12 (which is 36).
- **Four dozen** would be 4 × 12 (which is 48).
- ...and so on

There are a few special cases to be aware of:

- A special word is used for a dozen dozens (12 × 12) – this is called a gross. A gross means 144.
- Sometimes people refer to 6 items as a half dozen – because 6 is half of 12.
- Finally, if you ever hear "a baker's dozen", it doesn't actually mean 12 – it means 13!

Chapter 16: Squares & Tricks with Squares

We've covered all the times tables, so it's up to you to practise now.

The more that you practice, the better at times tables you will get! And the better that you get at times tables, the better you will get at math!

Before we finish though, I will introduce you to one last idea:

- If you multiply a number by itself, that is called "**squaring**". So, for example, "4 squared" means 4×4, which is of course 16.
- **We can indicate that we want to work out the square of a number by writing a little number 2 next to the number**. So, for example, instead of writing "4 squared", we normally would write 4^2. It still means 4×4, which is of course still 16.

Here is a list of all the squares:

- $1 \times 1 = 1$ squared $= 1^2 = 1$
- $2 \times 2 = 2$ squared $= 2^2 = 4$
- $3 \times 3 = 3$ squared $= 3^2 = 9$
- $4 \times 4 = 4$ squared $= 4^2 = 16$
- $5 \times 5 = 5$ squared $= 5^2 = 25$
- $6 \times 6 = 6$ squared $= 6^2 = 36$
- $7 \times 7 = 7$ squared $= 7^2 = 49$
- $8 \times 8 = 8$ squared $= 8^2 = 64$
- $9 \times 9 = 9$ squared $= 9^2 = 81$
- $10 \times 10 = 10$ squared $= 10^2 = 100$
- $11 \times 11 = 11$ squared $= 11^2 = 121$
- $12 \times 12 = 12$ squared $= 12^2 = 144$

Squares have many different uses in math, so it is definitely worth trying to remember them.

Squares can also help you with other multiplications: **If you want to multiply a pair of numbers that are 2 apart, then work out the square of the number that goes in between them, and then take 1 away**.

Here are some examples:

- If you wanted to calculate 5×7: You would notice that 5 and 7 are 2 apart (because $5 + 2 = 7$). The number that goes in between 5 and 7 is 6. 6 squared is 36, and we then take 1 away from 36 giving 35. So 35 is the answer to 5×7.
- If you wanted to calculate 8×6: You would notice that 8 and 6 are 2 apart (because $8 - 2 = 6$). The number that goes in between 8 and 6 is 7. 7 squared is 49, and we then take 1 away from 49 giving 48. So 48 is the answer to 8×6.

Chapter 17: Conclusion - All the Times Tables!

We've covered all the times tables, as well as lots of different ways to remember them – or at least o work them out.

We will finish with a table showing all the times tables! Remember to keep practicing!

If you enjoyed this book or it helped you, please post a positive review on Amazon!

Anyway, here's the table showing all the times tables that we've learned!

	0X	1X	2X	3X	4X	5X	6X	7X	8X	9X	10X	11X	12X
0X	0	0	0	0	0	0	0	0	0	0	0	0	0
1X	0	1	2	3	4	5	6	7	8	9	10	11	12
2X	0	2	4	6	8	10	12	14	16	18	20	22	24
3X	0	3	6	9	12	15	18	21	24	27	30	33	36
4X	0	4	8	12	16	20	24	28	32	36	40	44	48
5X	0	5	10	15	20	25	30	35	40	45	50	55	60
6X	0	6	12	18	24	30	36	42	48	54	60	66	72
7X	0	7	14	21	28	35	42	49	56	63	70	77	84
8X	0	8	16	24	32	40	48	56	64	72	80	88	96
9X	0	9	18	27	36	45	54	63	72	81	90	99	108
10X	0	10	20	30	40	50	60	70	80	90	100	110	120
11X	0	11	22	33	44	55	66	77	88	99	110	121	132
12X	0	12	24	36	48	60	72	84	96	108	120	132	144

For other resources to help you with your times tables, please go to:
http://www.suniltanna.com/multiplication

To find out about other educational books that I have written, please go to:

- For math books: http://www.suniltanna.com/math
- For science books: http://www.suniltanna.com/science

Remember: If you enjoyed this book or it helped you, please post a positive review on Amazon!

www.ingramcontent.com/pod-product-compliance
Lightning Source LLC
Chambersburg PA
CBHW041238200526

45159CB00031B/1866